It's Autumn...
Let's Learn!

An Early Learning Numeracy Book for Ages 3-5

By L. C. Benjamin

Copyright © 2026 L. C. Benjamin
Published by Benjamin's Press

All rights reserved.
No part of this book may be reproduced or shared without written permission from the publisher, except for brief quotations used in reviews.

First edition, 2026
Written by L. C. Benjamin
Illustrations created by L. C. Benjamin with assistance from AI tools.

Printed in Australia
ISBN: 978-1-7644551-6-9

For enquiries: BenjaminsPress2026@yahoo.com

About This Book

Children develop number sense gradually through everyday experiences. It's Autumm... Let's Learn! has been consfully designed to support early numeracy in a gentle and engaging way, using familiar autumn scenes and simple language.

As children move through the book, they practise counting, comparing groups noticing patterns, and exploring has numbed can change. The pages are arranged to build confidence first, then gradually introduce new ideas through observation, discussion, and play.

Created by a mathematics teacher, this book encourages children to talk about numbers, explian what they see, and make sense of quantities in ways that feel natural and meaniugful.

While this is a numeracy book, you may wish to extend learning by talking, drawing, or creatively responding to the autumn scenes together once counting is complete.

For Parents and Carers

This book is designed to build early number sense step by step, using familiar autumn scenes and simple, everyday language.

You can support your child by exploring each page together. Encourage your child to point, count aloud, and talk about what they notice. There is no need to rush. Some children may enjoy spending longer on certain pages.

How the book is organised:

• Core counting (pages 5-14):
Counting from 1 to 10 using clear, simple scenes.

• Reinforcement (pages 15-16):
Practising counting and recognising quantities.

• Comparing and grouping (pages 17-19):
Noticing more, fewer, same, and different.

• Early operations (pages 20-23):
Exploring adding and taking away through pictures.

• Reasoning and patterns (pages 24-27):
Sharing fairly, spotting patterns, and encouraging discussion.

Most importantly, keep the experience relaxed and enjoyable. Learning about numbers should feel natural, positive, and shared.

1

1 child plays in the leaves.

Let's count the child.

2

2 boots sit by the door.

Let's count them together.

3

3 pumpkins rest on the ground.

Let's count.

4

4 squirrels gather acorns.

Let's count the squirrels.

5

5 scarecrows stand in the field.
Can you count them?

6 acorns lie under the tree.
Count each acorn slowly.

7

7 raindrops fall from the sky.
Let's count together.

8

8 leaves are on the ground.

Count each leaf.

9

9 apples fill the basket.
Can you count them all?

10

10 autumn things are here.

Let's count from 1 to 10.

Let's count the squirrels again.

Can you find 2?

Can you find 5?

Point to the group with 4 scarecrows.

Point to the group with 6 apples.

This group has more corn cobs.

This group has fewer corn cobs.

Point to the group with more.

These raincoats are the same.

These raincoats are different.

Can you tell why?

Which group of hedgehogs is bigger?

Which group is smaller?

Here are 3 acorns.

2 more fall down.

How many acorns are there now?

A hen lays 7 eggs.

3 eggs are taken away.

How many eggs are left?

8 pears are shared between 2 friends.

1 for you.	1 for me.
1 for you.	1 for me.
1 for you.	1 for me.
1 for you.	1 for me.

How many pears does each friend have?

2 baskets of berries.

Which basket has more?

How do you know?

Boot, umbrella, boot, umbrella...

What comes next?

Can you try making your own pattern?

What number comes next?

1 2 3 4 ___

6 7 8 9 ___

Can you spot 3 leaves?

Can you spot 5 acorns?

You counted numbers.
You compared groups.
You shared fairly.
You explored patterns.
Well done!

Child's name: _____

Let's Keep Counting

Keep looking for numbers wherever you go.

There is always something new to count.

About the Author

L. C. Benjamin is an Australian school teacher with nearly twenty years of experience supporting young learners. As a teacher and a mum, she is passionate about helping children build confidence with numbers through everyday experiences.

She creates early learning books that focus on clear thinking, gentle progression, and joyful discovery. Her work is guided by the belief that strong number sense grows best when children are encouraged to observe, talk, and make meaning at their own pace.

It's Autumn…Let's Learn! reflects her love of teaching, nature, and thoughtful learning experiences that help children feel capable and curious.

Dedication

For my children, with love.
For all little learners, and the grown-ups who count the world with them.

www.ingramcontent.com/pod-product-compliance
Lightning Source LLC
LaVergne TN
LVHW072051060526
838200LV00061B/4713